SCARECROWS

SCARECROWS

by Avon Neal · photographed by Ann Parker

Barre Publishing
Barre, Massachusetts

Distributed by Crown Publishers, Inc.
New York

Copyright © 1978 by Avon Neal and Ann Parker

All rights reserved. No part of this publication may be reproduced, stored in a retrieval system, or transmitted, in any form or by any means, electronic, mechanical, photocopying, recording, or otherwise, without the written permission of the publisher. Inquiries should be addressed to Clarkson N. Potter, Inc., One Park Avenue, New York, N.Y. 10016.

Published simultaneously in Canada by General Publishing Company Limited

First edition

Printed in the United States of America

A NOTE ON THE TYPE

This book was set on the Mergenthaler Variable Input Phototypesetter (VIP) in Janson, a type thought to have been designed by the Dutchman Anton Janson, who was a practicing type founder in Leipzig during the years 1668-1687. However, it has been conclusively demonstrated that these types are actually the work of Nicholas Kis (1650-1702), a Hungarian, who most probably learned his trade from the master Dutch type founder Dirk Voskens. This type is an excellent example of the influential and sturdy Dutch types that prevailed in England up to the time William Caslon developed his own incomparable designs from them.

This book was composed by Publishers Phototype and printed and bound by Kingsport Press, Kingsport, Tennessee.

Typographic design by Hermann Strohbach
Editorial direction by Carol Southern
Production supervision by Michael Fragnito

Library of Congress Cataloging in Publication Data

Neal, Avon.
   SCARECROWS
     1. Folk art—United States.   2. Primitivism in art—United States.
     3. Scarecrows.   I. Parker, Ann   II. Title.
NK805.N372   779-9-745593   78-12868
ISBN 0-517-53500-9
ISBN 0-517-53501-7 pbk.

To our good friend
CORVUS AMERICANUS
Thanks for making this book
possible

Scare • crow (-krō), *n* 1. An object, usually suggesting a human figure, set up to frighten crows, etc., away from crops; hence, anything terrifying without danger.
    —Webster's New Collegiate Dictionary

Self-portrait of the scarecrow maker

Introduction

From man's earliest efforts to cultivate the soil he has relied on scary forms created in his own image to protect his crops. Long before Columbus set foot on the North American continent, Indians were erecting scarecrows to watch over their cornfields. Later, the Colonists used them extensively as they heeded the planters' adage and dropped four kernels into each hill of corn, "one for the birds, one for the worms, and two to grow!" More often than not dressed in the farmer's cast-off clothing, these intimidating minions stood guard while he tended to other chores.

In great-grandpa's day no field or backyard garden plot was complete without a scarecrow. They ranged from simple crossed-stick figures draped with old gunny sacks to sartorial masterpieces fit for display in the most elegant costume museums. Spring planting with all its hopes and cautions stimulated the creative urge and prompted raids on closets, barns, and musty attic trunks for an assortment of old clothes suitable for a scarecrow's attire. Anything from last season's fancy dress outfits down to faded overalls and shredded straw hats was fair game to the scarecrow enthusiast. Then, as now, the purpose was twofold—to frighten away unwelcome predators and to brighten an otherwise unadorned landscape.

By the time the nineteenth century rolled around, the lowly scarecrow was firmly established in our folklore and had earned a secure niche in the art and literature of this country. From the Bible to Hawthorne to the *Wizard of Oz* right up to modern-day children's books, they have been praised by poets, writers, and artists who have made them the subjects of endless paintings and illustrations.

A compelling air of magic and superstition lingers about these quaint figures. Our folklore is filled with tales of delinquent boys being frightened out of their wits when suddenly coming upon a particularly horrific example in a neighbor's watermelon patch and of tramps replenishing their wardrobes at the expense of helpless scarecrows, exchanging their threadbare coats and worn-out pants for more serviceable garments. During the late Victorian era, when fashion reigned supreme, it was not unusual for field effigies to spring full-blown from the previous season's outmoded Sunday best, and it was the lucky vagabond who plundered one in the dead of night. His booty might include a slightly frayed top hat as well as a frock coat and ill-fitting trousers, and little did he care if the switch turned him into one of those ridiculous characters portrayed in book illustrations of the period.

Today's tramp, however, would be hard put to come away with anything wearable from some of our contemporary field guardians. His booty would most likely consist of synthetic rags and double-knits plus the spin-offs from modern technology—strips of metal foil, garish reflectors, tin cans, plastic doodads, glass baubles, aluminum pie pans, and all things that flash and glitter in the sunlight or jangle nervously at the slightest breeze.

It is interesting to note who makes scarecrows and why. Women more often than men turn their artistic talents to the joyful task of scarecrow construction. Although the majority of their creations are male, it becomes increasingly apparent that female figures appear more and more frequently in present-day gardens. Children are the most spontaneous and ambitious of the scarecrow makers, while summer folk contribute more in the way of embellishment. Many of the finest scarecrows are put together by immigrants from Europe, generally older people who brought the custom with them from Germany, France, Italy, Portugal, Poland, or other countries steeped in this rural tradition.

When scarecrows were meant to be more utilitarian than ornamental, belief in their effectiveness was widespread throughout the community. Although many were highly elaborate, most were plain and practical. A tattered black coat flapping in the wind was enough to keep crows at bay, while a sweaty work shirt suspended from a fence post discouraged the nocturnal forays of hungry deer.

In this age of skepticism it is safe to say that practical people don't put much faith in scarecrows anymore. Of course, the very idea of the scarecrow's perennial success depends upon faith—faith in the maker's ability to contrive a convincing effigy with lifelike features, as well as faith that such a grotesque and frightening presence can deter birds and other garden pests from exacting their annual tribute from the farmer's labors. Most scarecrow builders believe implicitly in their creations and will defend them against all arguments, even when doubters can gaze beyond to a makeshift mannequin and see birds cheerfully nesting in the dilapidated haberdashery topping its fabricated head. That all-too-familiar spectacle is enough to betray any notion of infallibility, even among hard-core believers. Nevertheless, some stubborn spark of romanticism continues to flicker, for these are the young-in-heart troupers who perform the annual rites of spring by arranging old clothes over a gawky stick figure properly situated among their sprouting vegetable rows.

It is remarkable how many scarecrows end up looking like the people who make them. A portly farmer hangs a ragged denim jacket and a pair of triple-patched overalls on a post and stuffs them with straw until their seams split and their buttons pop; the lean, lanky type sticks a sparsely clothed, slim-jim figure in his garden. This phenomenon has often been sheepishly noted by passersby who have hailed a friend from a distance only to realize upon pulling alongside that they were addressing a newly constructed field warden. Such resemblances are no doubt coincidental, due more to the builder's having used his own discarded apparel than to any deliberate attempt at self-glorification. Perhaps it is appropriate that, as man was created in God's image, so scarecrows are created in the image of their makers, too.

Scarecrows are ephemeral creatures, rarely lasting beyond their season in the sun. Occasionally one is kept in a barn or shed to be refurbished for another spring planting, but most often they land on the rubbish heap, entwined in garlands of withered squash vines or stacked with brittle bean poles beside a garden wall. The skeleton of the once-imposing crop-watcher may eventually be broken into kindling lengths to warm the hearth some frosty evening when the harvest is in and conversation has turned to snowfalls, sledding, and Christmas dinner.

We are indebted to that dedicated band of traditionalists who have kept this art form alive; because of them, the common scarecrow has persisted, even proliferated, in modern times. After all has been said, the scarecrow remains the universal symbol of man's striving to wrest his living from the soil. As the farmer's loyal helpmate, this deserving sentry has flourished in all lands and throughout the ages, keeping vigil over field and garden from the dawn of agriculture right down to the present day. It is gratifying to know that, without pretense or fanfare, the scarecrow has in recent years been recognized as one of the truly naïve folk art forms. As the awareness of our artistic heritage broadens, so does the sophistication of folk art connoisseurs. Although it may be awhile before collectors actually install scarecrows as living room sculpture, examples of these whimsical figures have already been transported from their natural environment to major museums, for exhibition along with whirligigs, ships' figureheads, and cigar-store Indians. They are as fleeting as the seasons but, fortunately, the photographer's art enables us to explore and document their magical world, to capture their haunting beauty, and to preserve those delightful qualities that amuse both young and old alike.

By their very persistence, scarecrows have become the quintessence of rural life, recalling a bygone era when times seemed simpler and happier in the minds of those who seek pleasure in memory. Above all, scarecrows add color and humor to the fields and make them more friendly to man if not to birds.

Ann Parker and Avon Neal
Thistle Hill
North Brookfield, Massachusetts

Mop-bearded sentry guarding sunflowers

Headless French scarecrow

Shrouded figure in poppy field

Ecological Amazon

Crooked man with hat askew

Colossus in purple velveteen

16

Cornfield guardian with umbrella

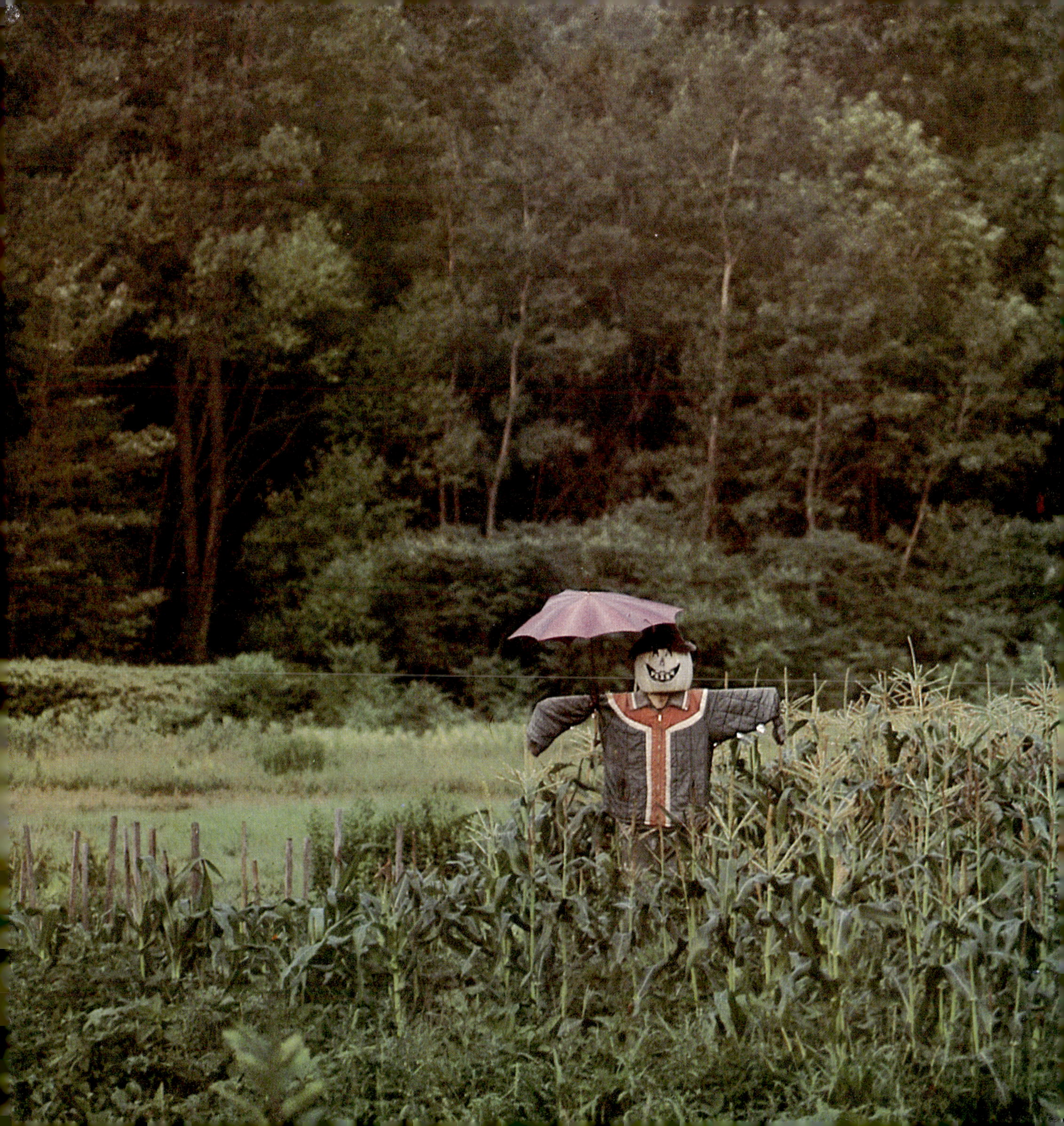

Barrel-chested varmint chaser

Bean-row dancer

Leaning lady in bucket bonnet

Vinyl-visaged sky watcher

Weathered rubber mask under gunshot hat

Traditional straw-stuffed mannequin sporting overalls

Cotton-headed field ghost

Clown-faced rustic

Gaunt specter in medieval garb

Portuguese scarecrow in cloudscape

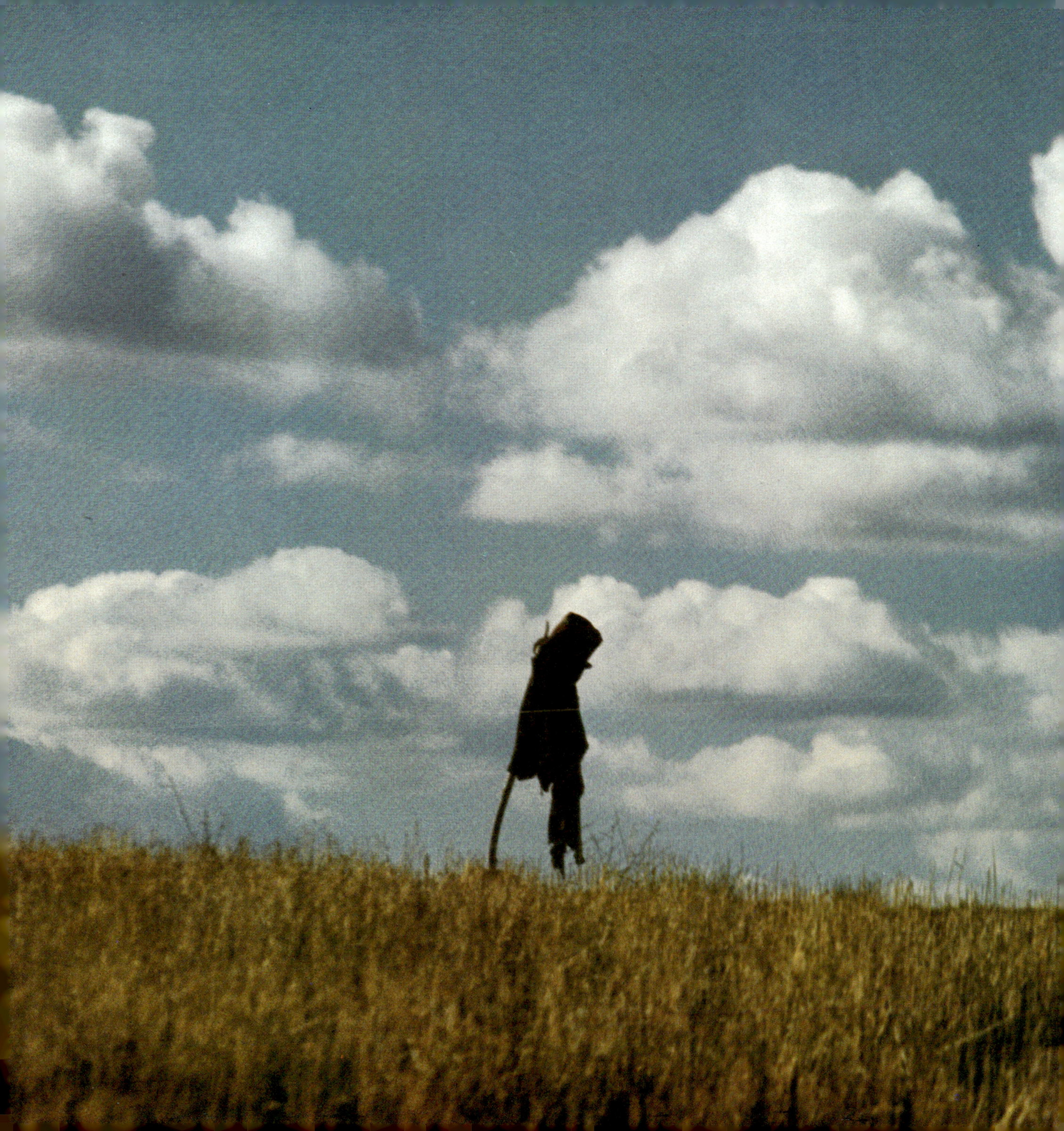

Tattered coat hung from apple tree

Modish pea-patch madam

Formidable garden granny

Scarecrow clothed in "wind-worried" rags

Simple post figure with flowerpot

Close-up of straw-helmeted field giant

Florid-faced ruffian

Turbaned watchman guards old shed

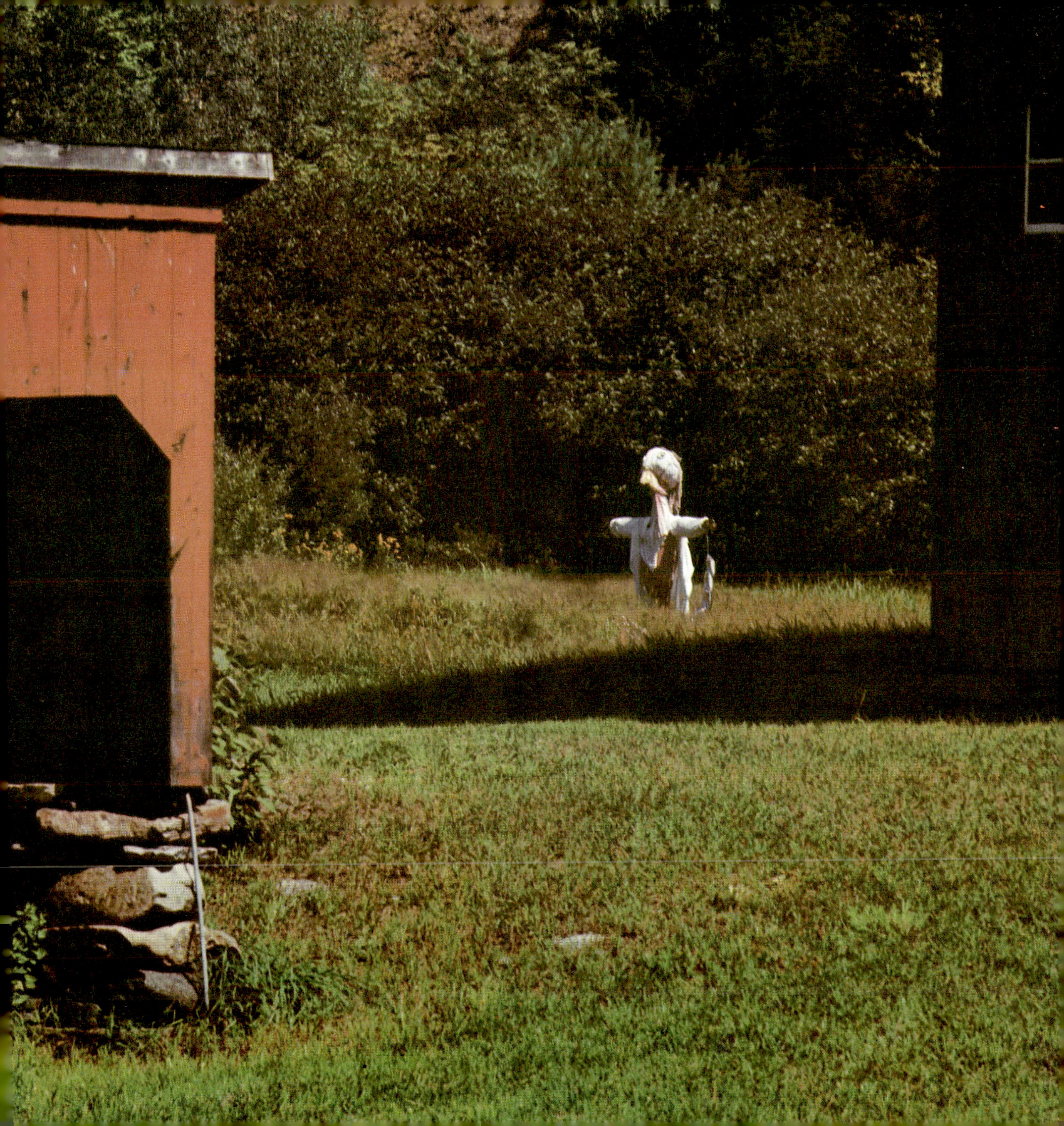

Friendly fellow with terry cloth head

Slim-jim

Headless trio amid flourishing greenery

Tin-can man with hubcap head

Beer-bellied overseer

Slouching debutante modeling blue denim

Free-swinging effigy guarding olive grove

Bespectacled straw head in truck-garden patch

Stick-wielding garden sorceress

Stern custodian of the vegetable plot

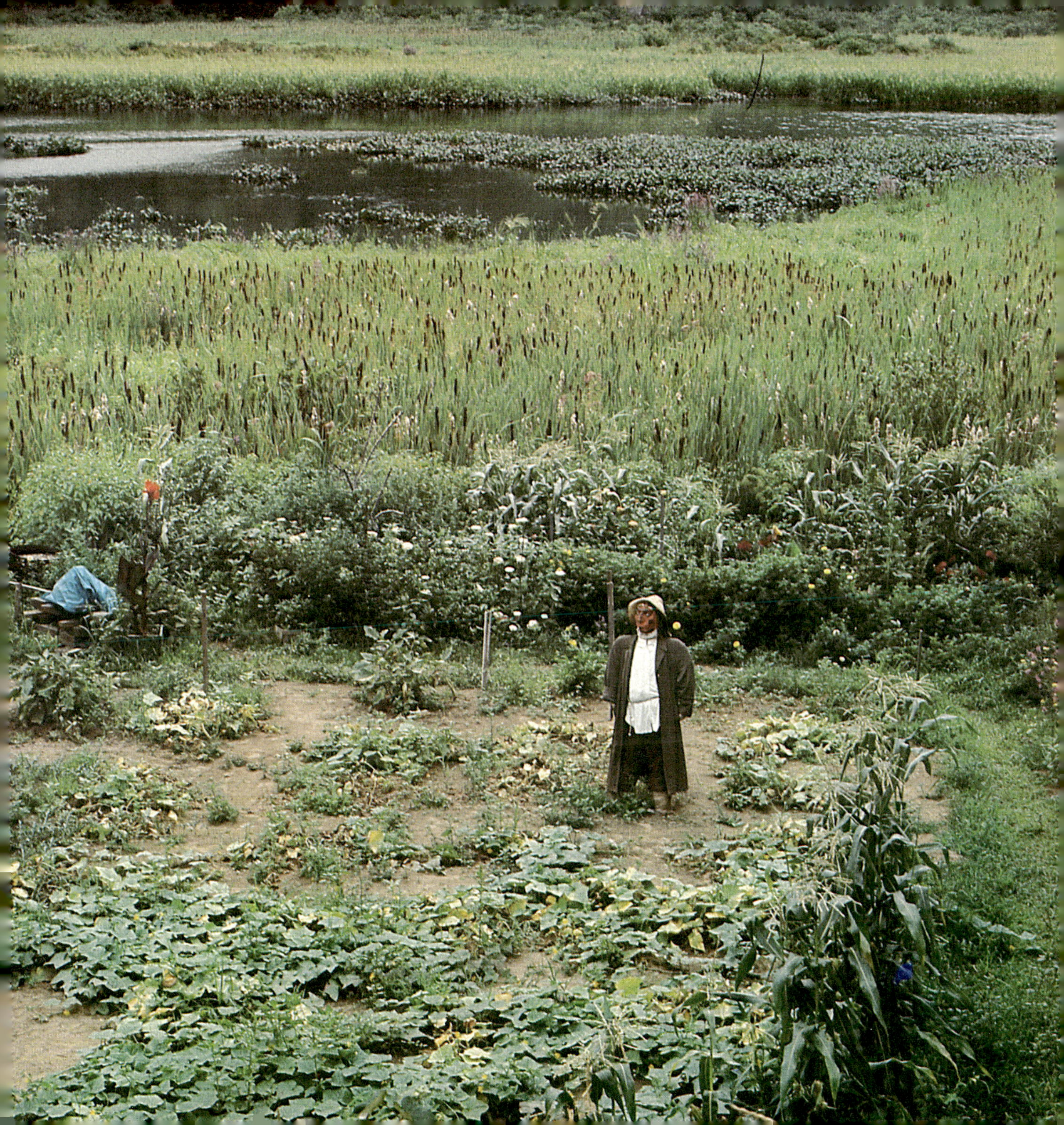

Sinister warning on French corncrib

Plebeian scarecrow basking in late sun

Extraterrestrial crop warden

Dainty stick figure welcoming spring

The garden dowager

Spring scene in Catalonia

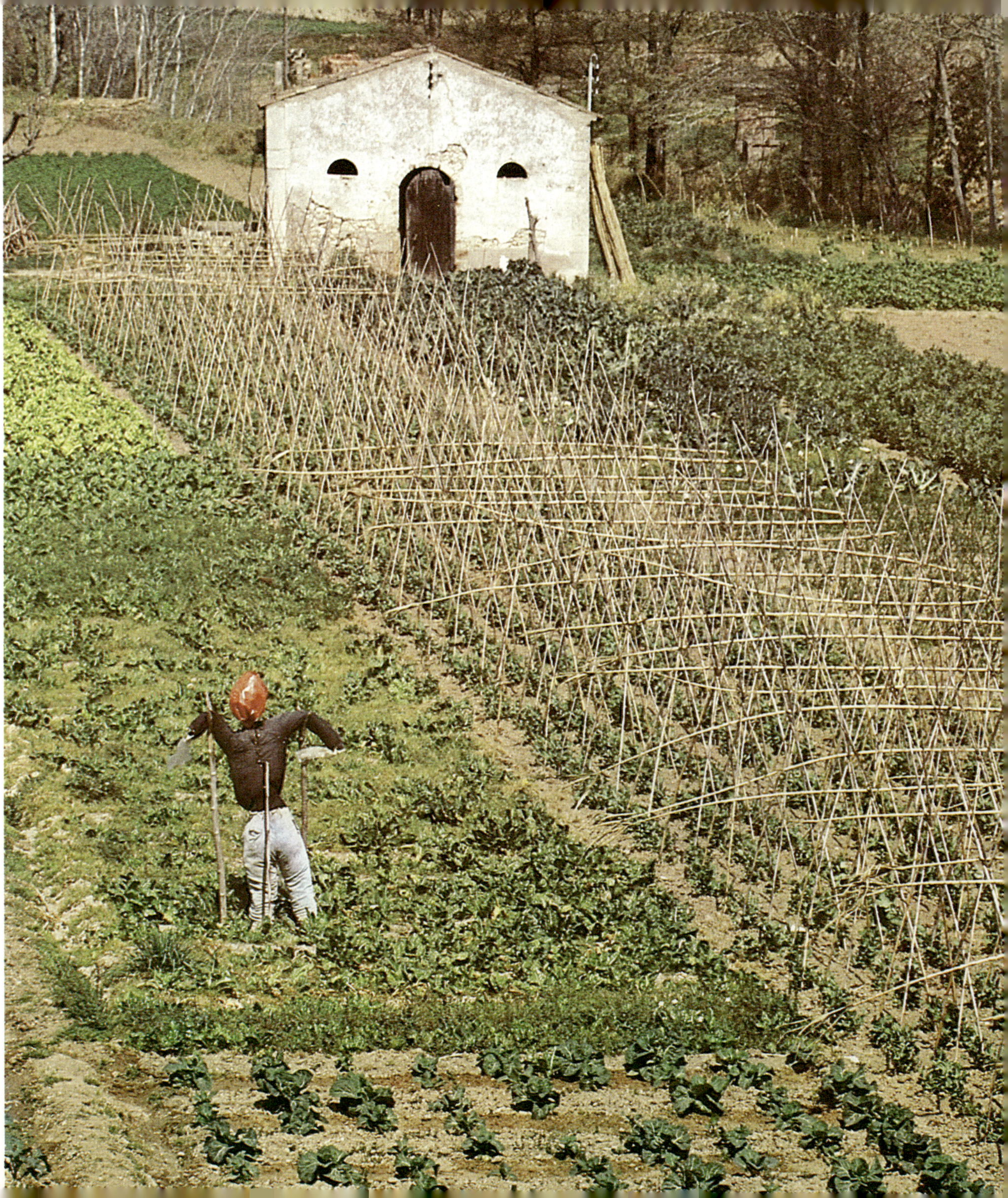

Cruciform figure with cotton shift

Ephemeral folk figure: Spring

Ephemeral folk figure: Summer

Mrs. Wiggs of the cabbage patch

Ephemeral folk figure: Autumn

Blithe spirit draped with plastic

Ephemeral folk figure: Winter

Skeletal relic of seasons past

Survivor after winter storm

Hard-hatted Japanese garden sentinel

Straw-helmeted field giant

Typical jolly farmer scarecrow

Vintage effigy in military uniform

Mannequin dressed from an attic trunk

Black-robed penitent

Old-fashioned scarecrow

Fur-coated crone and friend

Jezebel costumed in black vinyl raincoat

Dangling protector with cowbell

Ballet practice in a Portuguese vineyard

Peg-legged gatekeeper

Scarecrow wearing moth-eaten sweater

Paper-bag newcomer to berry patch

Cape Codder with lobster markers

Pie-panned sky-gazer

The bare-boned essence

Fair damsel holding parasol

Crow challenger

Loop-legged line-walker

Grotesque caretaker of field and barn

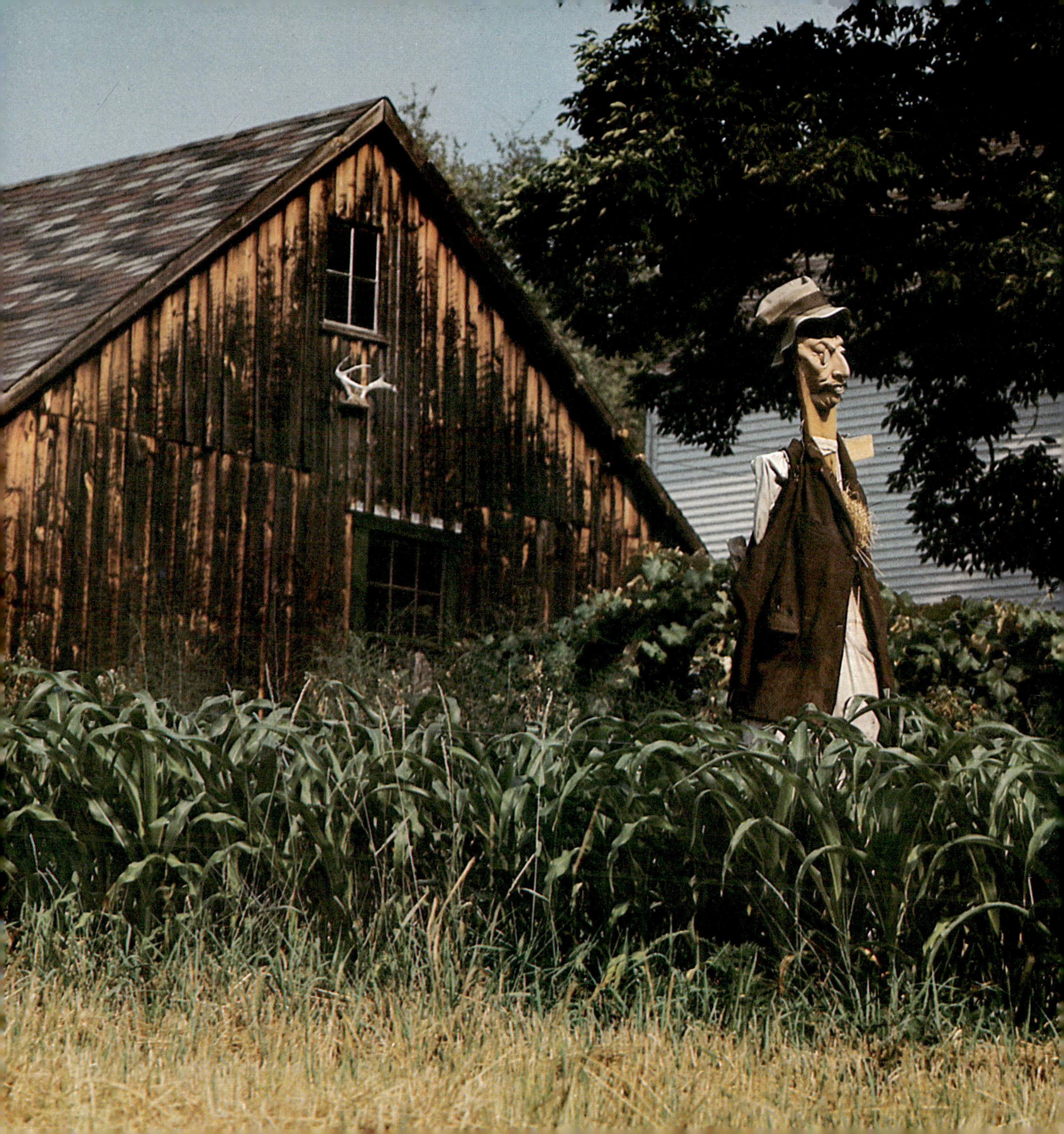

Landlocked mermaid

The Red Baron of Wilbraham

The tall weed stalker

Prim and proper lid dangler

Rope-bearded garden dude

Funny little fellow

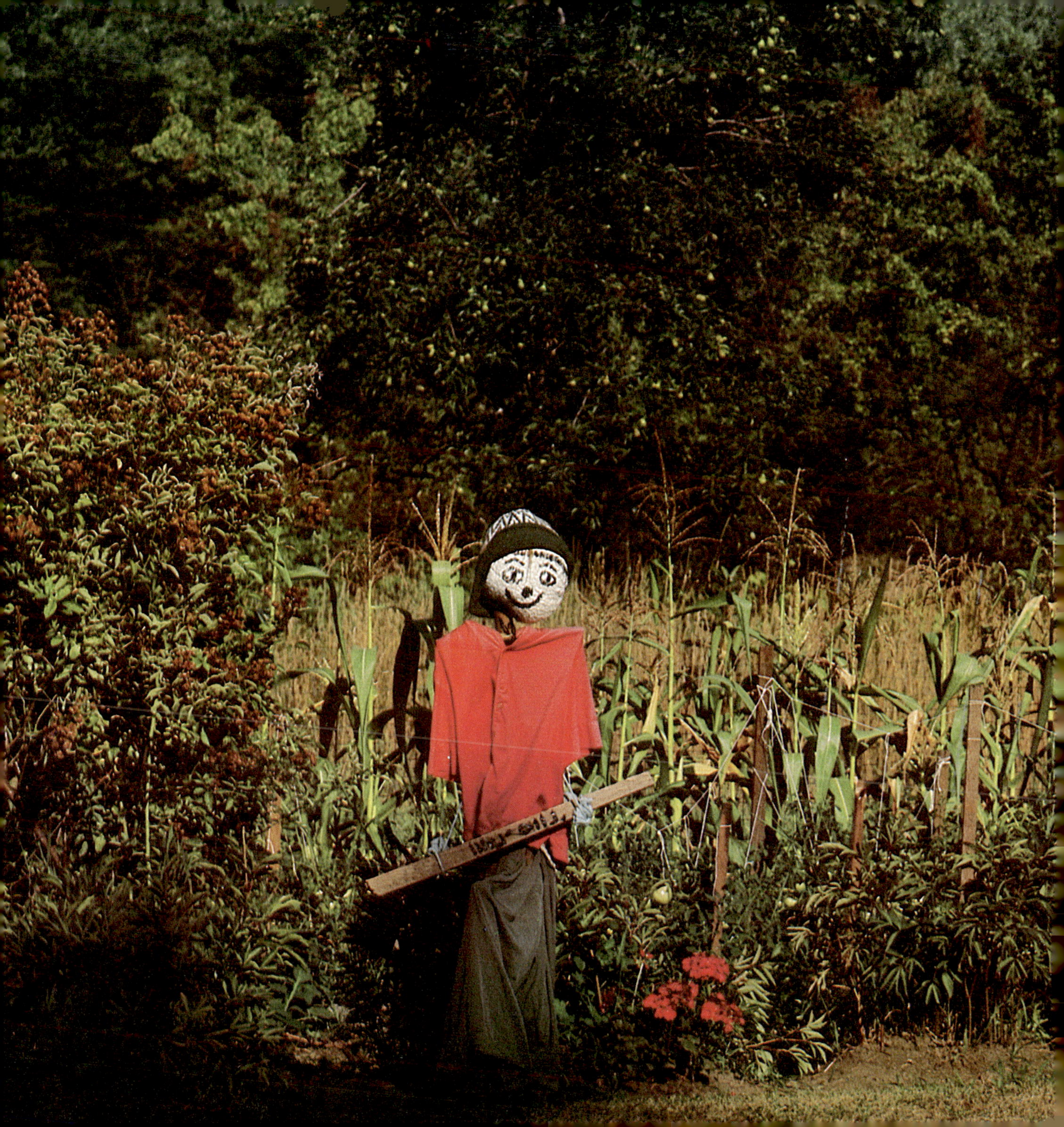

Fashionable lady in negligee

Fashionable lady viewed from window

Country bumpkin

Monumental field image

Snow-covered garden dude

Sartorial elegance in winterscape

The demure flower girl

Scarecrow in winter setting

ABOUT THE AUTHORS

Scarecrows is one of several endeavors Ann Parker and Avon Neal have undertaken together, exploring and documenting various aspects of folk art. They are best known for their extraordinary stone rubbings now found in many museum collections and for their highly praised book *Molas: Folk Art of the Cuna Indians*. Another book, *Ephemeral Folk Figures*, about scarecrows, snowmen, and harvest figures was published in 1969. The Neals are also contributors to such periodicals as *American Heritage*, *Art in America*, the *Smithsonian*, *Americana*, and *Life*.

   The Neals' fascination with scarecrows began quite by accident nearly twenty years ago while they were traveling the backroads of New England on a two-year research grant to study Early American gravestones. Since then, their pioneering studies in folk art have taken them into rural areas in many parts of the world, and on each trip new scarecrows are found and recorded. This fieldwork slowly evolved into a significant archive from which they have selected some of their favorites for this book.

   Scarecrows are ephemeral. Those presented here have already lived out their moment in the sun. Without Ann Parker's poetic photographs they would be lost to us forever.